AVIATION

Volume 1
Balloons and Airships

GW01182712

A Portrait in Old Picture Postcards

by

David B. Leeder

S.B. Publications
1991

To my wife, Adrienne, and daughters Corinna and Imogen

First published in 1991 by S.B. Publications,
Unit 2, The Old Station Yard, Pipe Gate,
Market Drayton, Shropshire, TF9 4HY

British Library Cataloguing in Publication Data

Leeder, David B.
  Aviation.
  Vol. 1: A portrait in old picture postcards.
  I. Title
  629.1309

ISBN 1 870708 90 3

Typeset, printed and bound by Manchester Free Press,
Paragon Mill, Jersey St., Manchester, M4 6FP. Tel: 061-236 8822

CONTENTS

Front Cover: The Deutschland, 1910.

BIBLIOGRAPHY

W.T. O'Dea (1966) *Aeronautica*, HMSO London
C.H. Gibbs-Smith (1966) *Aeronautics*, HMSO London
Christopher Elliott (1971) *Aeronauts and Aviators*, Dalton
M.J. Bernard Davy (1941) *Air Power and Civilization*, Allen and Unwin
Geoffrey Chamberlain (1984) *Airships Cardington*, Dalton
John Fabb (1980) *Flying and Ballooning*, Batsford
N.D.G. James (1983) *Gunners at Larkhill*, Gresham Books
John W.R. Taylor (1968) *Pictorial History of the R.A.F.*, Allan
Gordon Kinsey (1988) *Pulham Pigs*, Dalton
Robin Higham (1961) *The British Rigid Airship 1908-1931*, Foulis
T.E. Guttery (1973) *Zeppelin — An Illustrated Life*, Shire Publications
Kenneth Poolman (1960) *Zeppelins over England*, White Lion Publishers
Jane's Book of Airships, Collier Books (1977), Edited by Lord Ventry & Eugene Kolesnik

ACKNOWLEDGEMENTS

The author wishes to thank the Aero Philatelic Club of London for permission to reproduce the postcard on page 1.

Every effort has been made to obtain, where necessary, permission to publish these cards. All the postcards in this Volume 1 are from the collection of the author.

Editing and proof-reading: Frank Rhodes of Lightwood.

Sales, marketing and publishing: Steve Benz.

INTRODUCTION

Since the dawn of time men have dreamed of being able to fly, and in this book I hope to show through postcards some of the first men who braved the elements with their balloons and airships.

It was on 15 October 1783 that Pilatre de Rozier and Marquis D'Arlandes in a Montgolfier balloon made the first manned flight. From that day on men and women all over the world became fascinated at the thought of travelling through the air in balloons. When Benjamin Franklin was asked in Paris, 'What is the use of balloons?' he replied, 'What is the use of a new-born infant?' Many methods were used in the construction and propulsion of the balloons. Also in 1783 in Paris, the Robert brothers were the first to use hydrogen gas. In 1821, Charles Green used coal gas for his ascent. Jean Pierre Blanchard at this time experimented with oars to propel his way across the sky, and many others were building and flying 'their machines of flight'.

The French were the first to use the balloon for military observation, at Mauberge in 1794, and in 1849 the Austrians used balloons to drop bombs on Venice. During the American Civil War they were used for artillery spotting, and during the Siege of Paris, in 1870/1871, as a means of escape. Balloon Sections were formed to accompany military expeditions to Bechuanaland and the Sudan in 1884 and 1885. Then in 1887 the British Government recognised the huge potential of balloons and made a Major Templer 'Instructor in Ballooning'. In 1890 a Balloon Section was added to the Royal Engineers. During the Boer War the British spied on the Boers with three balloon sections raised by the army.

As balloons became more reliable and engines were attached to the undersides of the envelope, non-rigid airships came into being. In France, Santos-Dumont was experimenting with his dirigibles, and in Britain E.T. Willows was doing the same with his Series 1 to 6. In 1907 the Balloon Section at Farnborough made the Nulli Secundus, a non-rigid airship powered by a 50 h.p. engine. In Germany a professional soldier was building airships — Count Zeppelin.

At this point something should be said about collecting postcards of this period. The two main types are the artistic impression cards (before postcards were first produced in 1894) and the photographic cards produced by the local 'on the spot' photographer and the official pictures. The condition of all such cards is important; many photographic ones have deteriorated over the years because of the impermanent nature of printing fixatives used in the early period of photography. The artistic cards are usually in better condition but have the disadvantage of not always being accurate in detail. This is when the collector has to research that extra guide-rope, rudder or if lucky, an odd number or letter on the fabric of the balloon or airship.

During the First World War both artistic and photographic cards were published by all the warring factions.

Many of the celebrities connected with this period were shown on cards and sold to the public as 'heros' and 'men of the moment'.

After the momentous years 1914/1918, the airships found a commercial use, and they sailed the continents and oceans until the R101 and Hindenburg disasters in 1930 and 1937 brought to an end the passenger service that had been a huge success up to that time, both for the passengers and the airlines. These two events had made people sit up and think about the safety of such a means of transport.

Now the hot-air balloons have made a come-back, and meetings are held countrywide. The multi-coloured envelopes, some depicting advertisements, make a colourful sky-scene. Airships in the form of advertising platforms have been seen recently — trying to make a commercial return.

Will we ever see the return of the giant, fare-paying, ocean-plying airships?

MONTGOLFIER BALLOON

This postcard, although modern, gives a good idea of the Montgolfier brothers' balloon being demonstrated in Annonay, France, in 1783. It was constructed of cloth lined with paper; the circumference was 110 feet. A wooden frame, sixteen feet square, held it fixed at the bottom. Their balloon stayed in the air for ten minutes, flown by De Rozier and D'Arlandes. It was very colourful, blue and gold, the sun rays with the face in the centre makes the balloon very distinguishable.

APEX 1983 B.P.E.

60th Anniversary of the Aero Philatelic Club, London
25th Anniversary of the British Airmail Society

1

UHLANS PURSUING A BALLOON.

SIEGE OF PARIS, 1870

Paris was besieged by the Prussians. Nothing was allowed in or out of the city. The cornered citizens got messages out of Paris by balloons manufactured especially for the purpose, and during the siege some 65 escaped, carrying 164 passengers, 381 pigeons, 5 dogs and eleven tons of mail. It is interesting to note that after the Franco-Prussian War, Zeppelin was appointed to command a squadron of these Uhlans.

2

GLASGOW INTERNATIONAL EXHIBITION

This is not exactly an advertising card, but clearly shows part of the Glasgow International Exhibition of 1901. The balloon would have been one of several that took part in the races organised during the Exhibition. Rides were also given to the daring public. The rope trailing beneath the basket was used as an anchor when landing.

LIFEBOAT SATURDAY, 1902

In order to raise funds the Royal National Lifeboat Institution staged balloon flights at Manchester for Lifeboat Saturday, the first on 20 September 1902. This card is obviously not one of those carried in the balloons, but was a facsimile of the 4,000 carried on the flight, which was manned by M. Gaudron and Dr F.A. Barton, and flown from the Botanical Gardens, Old Trafford.

LIFEBOAT SATURDAY, 1903

Again in 1903, the Royal Lifeboat Institution had another balloon flight. This flight, however, had to be abandoned when the balloon became entangled in a tree, and the postcards being carried had to be dropped a week later from a balloon ascent at a military tattoo in the grounds of Alexandra Palace. This particular card (a copy) was sold at auction in 1980 and made £880. The famous Jeannie Bowden signed the card.

HARBORD CUP, 1907

A massed start at Ranelagh for the annual race. The winning-post was Goring Railway Station, Sussex. This was a very popular event and drew many people to see the start. Because everyone wore their best dresses and hats, it has been compared with the races on Derby Day — another fashionable display.

6

Engineers Ballooning,
Long Valley, Aldershot

ROYAL ENGINEERS BALLOON SECTION

The first use of balloons for military observation was by the French army at Maubeuge in 1794. The British army employed them on active service in Bechuanaland and Sudan in 1885, and later in the Boer War. It was not until 1890 that the first Balloon Section was formed. In 1891 it moved to Aldershot. The Balloon Factory incorporating the School of Ballooning was established in 1892 under Major L.R.B. Templer as 'Instructor of Ballooning'.

ROYAL ENGINEERS. BALLOON SECTION,
PREPARING FOR AN ASCENT WITH
BALLOON 10,000 FEET CAPACITY.

Copyright

ROYAL ENGINEERS BALLOON SECTION

In 1909 the Factory and the School became separate establishments. In April 1911 the Balloon School became the Air Battalion, Royal Engineers Aldershot. Both cards show clearly the observer climbing the rigging into his basket. Gas cylinders can be seen on the horse-drawn carriages.

BEST WISHES

Large numbers of greetings cards were published depicting balloons during the public's enthusiasm in the early 1900s. This embossed balloon covered in forget-me-nots was sent from Swansea to Bath on 25 December 1909.

BEST WISHES

Another popular flower on greetings cards was the violet. Here we have two doves as crew, and the surfaces of the airship and gondola covered in violets. It was posted from Coventry to a local address on 5 June 1907.

Nulli Secundus

NULLI SECUNDUS

After a government cash injection for the Balloon Factory at Farnborough, the Nulli Secundus was launched in 1907. She was 122 feet in length and powered by a single Antoinette engine of 50 h.p. Its maximum speed was 16-18 m.p.h. Its first flight was at Farnborough on 10 September 1907. On the 5 October, with Colonel Capper and Samuel Cody, it flew round St Paul's in London, then landed at Crystal Palace, where it stayed for five days — marooned and deflated. In 1908 it was reconstructed and called Nulli Secundus II. It flew one flight of 18 minutes — and was then scrapped.

SHREWSBURY FLOWER SHOW
A balloon ascent was a popular item in the programme of many Edwardian floral fêtes, agricultural shows and similar open-air events. This card shows a balloon at the Shrewsbury Flower Show in 1910, surrounded by a large crowd. Note the sacks of sand used as weights before the actual ascent. The card was sent from Shrewsbury to Malmesbury on 17 August.

PIN-UP ALOFT

Music Hall artistes had postcards published for their fans. This one shows Billie Burke and Maie Ash over London, the basket superimposed over the Law Courts.

4069 N ROTARY PHOTO. E.C. MISS CARRIE MOORE. FOULSHAM & BANFIELD

PIN-UPS ALOFT

Another Edwardian actress aloft in a balloon basket. Miss Carrie Moore, with raised hand, peers down from a genuine basket, complete with the ballast sacks and anchor. Posted from Worcester to a local address on 4 January 1907.

Rhyl — The air here is excellent.

DEUTSCHS AIRSHIP "LA VILLE DE PARIS".

LA VILLE DE PARIS

The airship belonged to M. Deutsch de la Meurthe. It was powered by a 70 h.p. Chenu engine. At 201 feet in length it was the largest airship at this time. She was launched in 1907, and presented to the French Army where it was stationed at Verdun. At its stern were fitted 8 supplementary gas envelopes to assist its steadiness. This card was posted from Rhyl to Halesowen on 16 September 1909. Part of the message reads, 'Shall you get in this airship and fly to me Mary? Fond love Arthur.'

4 ° Dirigible airship "Clément-Bayard". — LL.

CLÉMENT-BAYARD

This airship from the Maison Clément-Bayard was built in 1908 and known as the C B 1. Its six ugly thick stabilising fins at the stern of the envelope and cagework to prevent the propeller blade striking the gas-bag in case of fracture are clearly seen. It was powered by two four-cylinder motors, driving a large single wooden propeller. The envelope was made by Astra, whose shed the airship is seen leaving. It was purchased by the Russian Army and was lost on trials in 1909.

AIRSHIP BETA

May 1910: Beta 1 was the name of the lengthened 'Baby' — now 104 feet. By July, with improved speed, she flew from Farnborough to London in 3 hours 45 minutes. In 1911 the Army fitted her with the first experimental wireless equipment in any airship in Britain. It was also the first non-rigid airship in the world to be moored to a mast, and to be inspected by King George V and Queen Mary. In the autumn of 1914 it patrolled London, and later that year was deflated to make way for the next Army airship.

PARSEVAL

This type of airship was made in Germany from 1906, to the original design of Major von Parseval. Of the 27 built, several found their way to other countries — Japan, Russia, Turkey, Italy and Great Britain. The Parseval had two ballonets inside the envelope, one fore, the other aft. This was so that the trip could be controlled. The trailing cables were used to enable the axis of the car to be kept horizontal. Most Parsevals were used for military purposes, and production of this type ceased in 1917.

PARSEVAL

In 1913 a German Parseval was tested and became Naval Airship 4. This Parseval and a French Astra-Torres were the only British naval airships, and were used to escort the original British Expeditionary Force to France. They were also used on patrol duties. The Parseval was 300 feet long and powered by 2 Maybach engines. Its maximum speed was 42 m.p.h. This card shows the Parseval on patrol over Dover harbour.

PARSEVAL

Parseval No. 4 made its maiden flight on 30 June 1913. After circling the Houses of Parliament and St Paul's, she returned to Farnborough. She patrolled Thamesmouth during August 1914, becoming the first British airship to operate in the war. She was used to train many crews during the next three years, but by 1917 her streamlined shape became distorted and she was deflated on 17 July.

FRANCO-BRITISH EXHIBITION

Advertising the Exhibition held in 1908 at the White City, London, below the balloon is a plan view of part of the site. Posted from Surbiton and addressed to St Thomas's Hospital, London on 13 July 1908. Part of the message reads, 'What do you think of this take-off of the Exhibition, I should certainly like to go there but not in this style, what say you.'

ASTRA TORRES

Seen here moored at Farnborough, she was launched in the spring of 1913, then flown from the Astra Company at Issy, France. At Farnborough she first flew in public on 12 June. After some mishaps she next appeared on 8 September. At the end of September she was handed over to the Royal Navy, and was used early in the war for photo-reconnaissance over Ostend and Zeebrugge. She was the last of the pre-war airships programme to be completed in time for combat use in 1914. In 1915 she was deflated and stored. The Astra Torres was 248 feet in length.

OBSERVATION BALLOONS

During the 1914-1918 war, the shape of the observation balloons changed. The 'sausage' type was built with large extra gas bags used as stabilisers. A large balloon with greater capacity meant higher altitude and better observation. The observer would be wrapped in heavy clothing as protection. Horses have now been replaced by heavy motor lorries, which were used for transportation and as a winch.

GERMAN OBSERVATION BALLOON

A British 'Spad' fighter attacking a German observation balloon; the observer can be seen jumping from his basket with his parachute. Not all observers had time to use their 'chutes, many being engulfed by burning fabric, and others being too close to the ground.

AIRSHIP ON CONVOY DUTY

CONVOY PATROL

During 1917 a renewed campaign against Allied shipping began, so a complex system of anti-submarine patrols was set up by the Royal Naval Air Service. Such patrols would be carried out by the new airships — SS Zero non-rigids. This postcard was posted at Epsom on 13 October 1918. Part of the message read, 'This is just how we used to see the Air-ships at Ventnor in July.' This card was a special War Bond Campaign card.

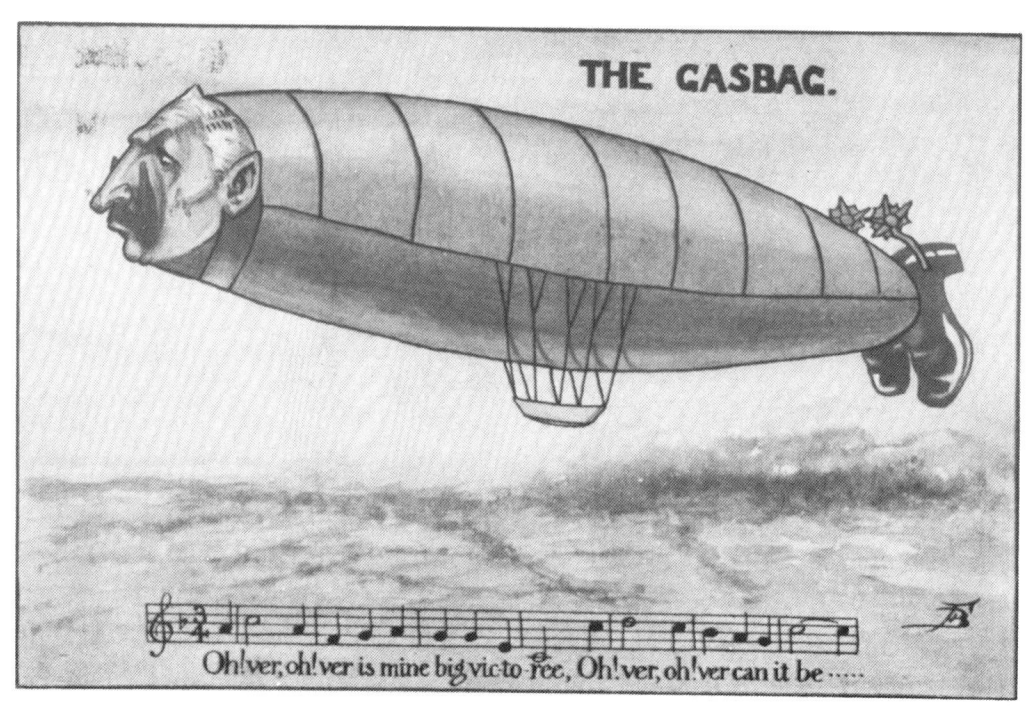

THE KAISER — THE GASBAG

Kaiser Wilhelm II depicted as The Gasbag, and from his tune, having doubts about his future and his Empire. Posted from Weston-super-Mare to Whitby on 3 August 1915, part of the message reads, 'Keep an eye for the chap on the other side he might come to see you, one never knows.' In fact two Zeppelins did raid the east coast on 12 August, killing and injuring 30 people and destroying 14 houses.

DEUTSCHLAND

Launched in 1910, the Deutschland was one of a fleet of commercial Zeppelins flying all over Germany. Over 37,000 Germans had bought tickets at branches of the Hamburg-Amerika Shipping Line. They had flown 100,000 miles, totalling 1,600 flights taking 3,200 hours. The Deutschland was 485 feet long and powered by 3 Daimler engines each of 120 h.p. Its maximum speed was 37 m.p.h. It was wrecked in a gale at Teutoburger Wald while carrying 20 passengers on the 28 June 1910.

DEUTSCHLAND
A greetings card, depicting the Deutschland airship. Sent from Chacewater to Redruth on 4 August 1910, this card was printed in Saxony.

LZ8 ERSATZ DEUTSCHLAND

So called because this Zeppelin replaced the LZ7 Deutschland which was destroyed in a gale in June 1910. The above Zeppelin was much the same in general design, but 7 feet longer and 3 feet extra in its diameter. Whilst mooring at its hanger a gust of wind snatched the airship from the hands of the 'helpers' and it was impaled on the wind shelters at the entrance to its hanger. This happened on 16 May 1911, but there were no casualties, many being rescued by fire escape.

Graf Zeppelins lenkbares
Luftschiff in voller Fahrt

SCHWABEN

Although a drawn postcard, the shape of the unidentified airship could be the Schwaben. Completed in 1911, in eleven months she had carried 1,533 passengers on 218 flights. She was destroyed by fire in a gale in June 1912. This card shows the Schwaben flying over an unidentified town, somewhere in Germany — with an enthusiastic population waving, and one lady in bottom right with binoculars getting that extra view.

Z III ZEPPELIN

The Z III Zeppelin was the sister ship of the Schwaben. Both of these were kept at Dusseldorf. They were both 459 feet long and powered by 3 Maybach engines, and the maximum speed was 47 m.p.h. The Z III, seen here rising over the town, was launched in 1912 and taken out of service in 1914.

Das Zeppelin'sche Luftsch ff über dem Brandenburger Tor

VIKTORIA LUISE

Germans looking with pride at the Viktoria Luise as she flies over the Brandenburg Gate in Berlin during 1912. Launched in 1912, she was one of the Deutsche Luftschiffahrts-Aktien-Gesellschaft fleet. The German Navy's first airship crews were trained aboard this and the Hansa, often flying then while passengers were aboard. The Viktoria Luise was 485 feet long, powered by 3 Maybach engines with a maximum speed of 49 m.p.h.. She was wrecked while docking on 1 October 1915.

VIKTORIA LUISE

This postcard shows the giant airship flying over Cologne, against a backdrop of the cathedral and the Rhine. Viktoria Luise and others in the fleet made many trips over the cities of Germany.

Friedrichshafen a/B. Landung des Luftschiffes „Graf Zeppelin" vom Flugzeug aus gesehen

FRIEDRICHSHAFEN

Friedrichshafen on Lake Constance was the home of Count Zeppelin's airships; here in 1900 he built a floating shed for his early experiments. Later, in 1914, allied pilots made a daring raid to attack the sheds. Right up to the late thirties Friedrichshafen was the home of the Graf Zeppelin I and II. Graf Zeppelin II was the last airship to be built at this base. Hitler ordered her to be broken up during 1940.

NIGHT ATTACKERS

Very much an artists impression; a Zeppelin completely illuminated by four searchlights, and four bursts of artillery gun-fire are near-misses. This card was published by permission of the War Office.

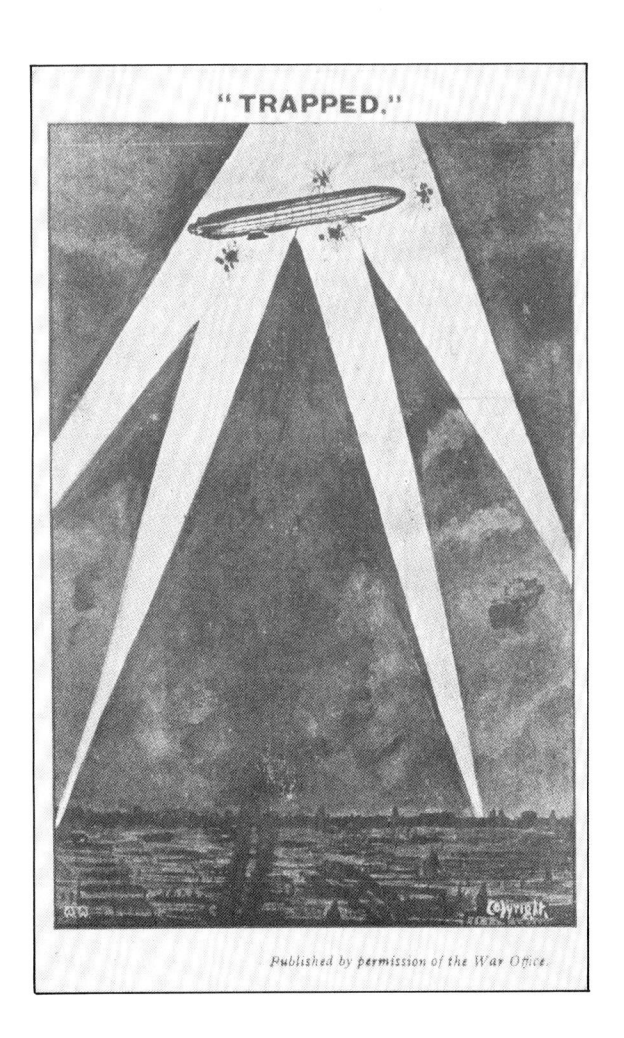

"TRAPPED."

Published by permission of the War Office.

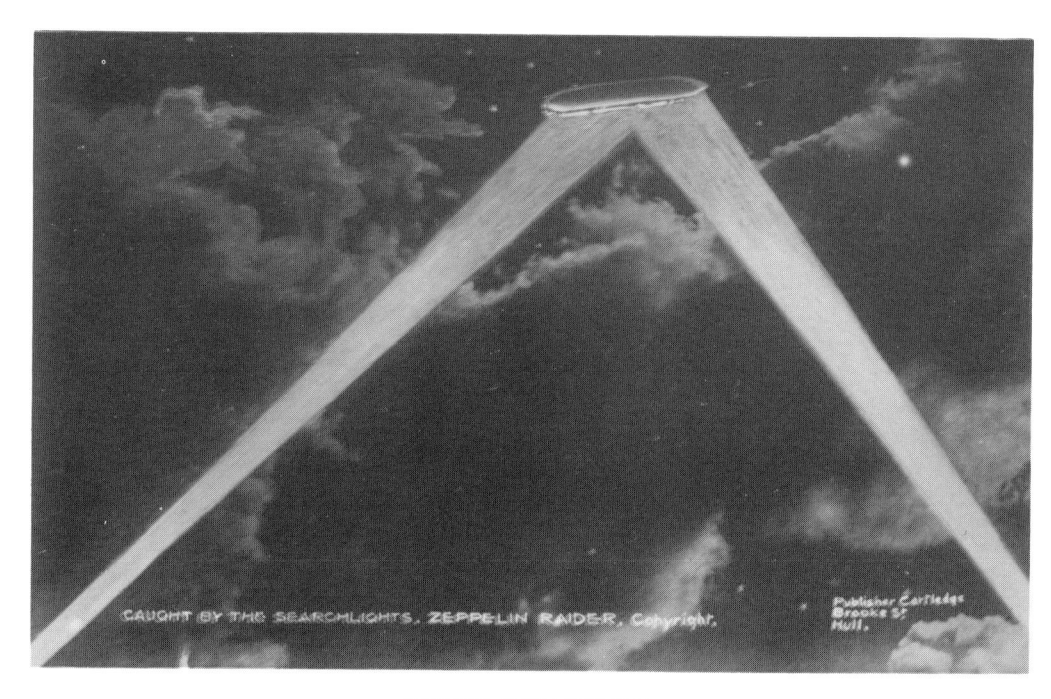

CAUGHT BY THE SEARCHLIGHTS. ZEPPELIN RAIDER. Copyright.

Publisher Cartledge
Brooke St
Hull.

NIGHT ATTACKERS

This card published by Cartledge of Brooke Street, Hull could have been on sale after two Zeppelins had raided the town on 5 March 1916. Seventeen were killed and fifty-two injured. Nearly all the damage was to civilian property. Both raiders, the L II and L 14, returned safely to their bases.

NIGHT ATTACKERS

Many cards were published showing airship raids, where the searchlights and airship were drawn in later. This card was posted on 4 August 1916 from Kirk Ella to Louth, on the east coast — both places in the path of raiding Zeppelins. Two days before this card was posted there had been a raid by six Zeppelins, and on the 8th, nine more airships had raided targets in Norfolk and Northumberland.

THE RAIDER
PUBLICATION SANCTIONED BY OFFICIAL PRESS BUREAU.

PUBLISHING OFFICE:
39, ST. ANDREW'S HILL, E.C.   (Copyright

NIGHT ATTACKERS

Like the previous card showing the silhouette against the church and chimney, this card shows the shape of a soldier looking up at the 'Zepp'. Although not numbered, these silhouette 'night raid' cards are very collectable. All of them are marked, 'Sanctioned by the Official Press Bureau.'

" ZEPP."
SANCTIONED BY THE OFFICIAL PRESS BUREAU.
Publishing Office:
39. St. Andrew's Hill, E.C
Copyright
H. Scott Orr.

Marine-Luftschiff L2 kurz vor der ersten Landung in Johannisthal am 20.9.1913.

ZEPPELIN L 2

Commercial Zeppelins were flying from sheds erected at Dresden, Baden-Baden, Hamburg, Frankfurt and Dusseldorf, and also at Johannisthal, where we see the L 2 landing on 20 September 1913. She had an internal keel and separate control car — and was 518 feet in length. On 17 October 1913 she was destroyed in the air by fire.

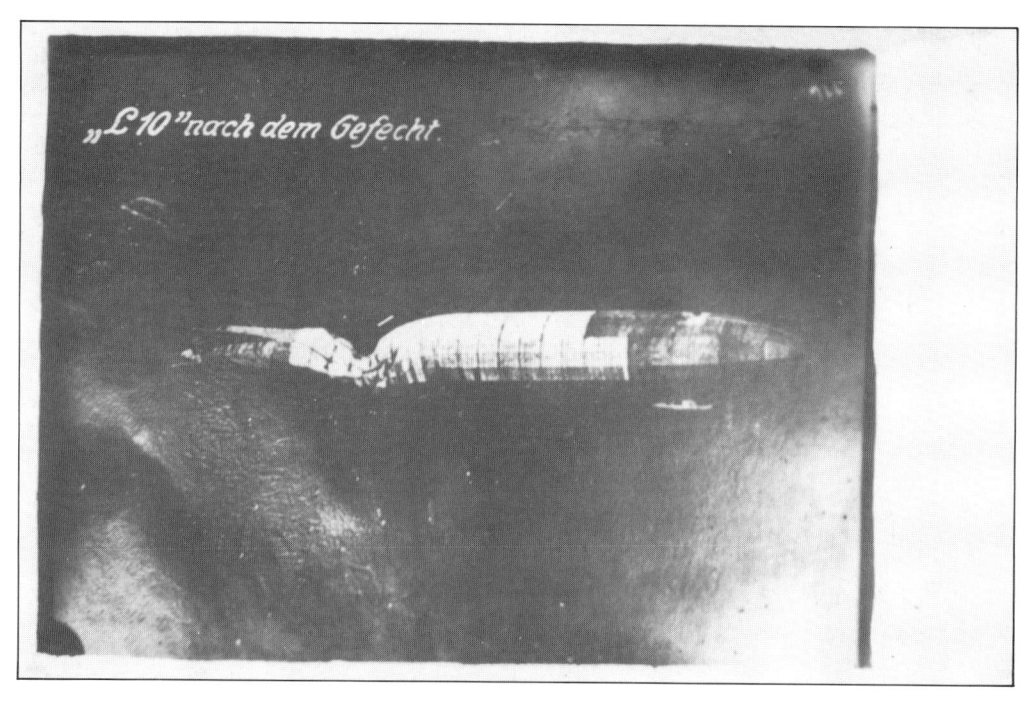

„L 10" nach dem Gefecht.

L 10 ZEPPELIN

'L 10 after the fight' reads this card. On 17 August 1915, three Zeppelins left their base and headed for England. One of them, the L10 captained by Fridrich Wenke, bombed Leyton, London, killing 10 people and injuring 48. Launched in 1915 it had a very short career. After the raid on Leyton, just under one month later, on 3 September, it was struck by lightning and destroyed. The card was published in Germany.

ZEPP RAID AS SEEN "SOMEWHERE WITHIN THE LONDON AREA," Autumn, 1915.

L 15 ZEPPELIN

The new Zeppelin L 15, with four others, made for London on 13 October 1915. It was reserving its bombs until well over the city. At 9.25pm the first bombs fell on 'theatreland.' Seventeen were killed near the Lyceum, and twelve badly wounded. The Gaiety theatre also had casualties. The Zeppelin went over the West India Dock to Limehouse. Here guns at Woolwich forced her to turn north for her home base.

Zeppelin Brought Down Off East Coast, Nov. 28. 1916.

L 21 ZEPPELIN

On 28 November 1916 ten Zeppelins raided northern England. As dawn broke the L 21, nine miles off Lowestoft, was overtaken by Flight Lieutenant Cadbury, Flight Sub-Lieutenant Fane and Flight Sub-Lieutenant Pulling of the Royal Naval Air Service. All three pilots shot at the L 21, and it was Pulling who fired fatal shots, bringing down the Zeppelin into the sea. Although not postally used, the message on reverse reads, 'As seen from Gorleston Cliffs.'

L 31 ZEPPELIN

The first of October 1916 found the L 31, captained by Heinrich Mathy, making for London, but because of accurate gunfire he had to abandon the city as a target, dropping most of his bombs on Cheshunt. Meanwhile, Second Lieutenant Tempest was patrolling over south-west London at 14,000 feet. Mathy and Tempest spotted each other and the chase was on. The Zeppelin was shot down at Potters Bar, and for this action Tempest was awarded the DSO.

"THE FOURTH"!!!!
Super-Zeppelin brought down in Flames at Potters Bar, Oct. 1st, 1916.
Reproduced by permission of "THE DAILY SKETCH."

THE STRAFED ZEPP 17th June 1917. No 3. Published by J. S. Waddell, Photographer, Leiston

L 48 ZEPPELIN

On 17 June 1917 four Zeppelins made for England. Before the coast was reached two had turned back. The third bombed Ramsgate and got clean away. The fourth, the L 48, steered towards the Suffolk coast. It dropped bombs near Martlesham. At the Orfordness Experimental station the alarm was raised and three pilots took off to intercept.

THE STRAFED ZEPP. 17th June, 1917. No. 5   PUBLISHED BY J. S. WADDELL, LEISTON

L 48 ZEPPELIN

Less than one month before this raid, the L 48 had been commissioned at Friedrichshafen. It was 644 feet long, and powered by 5 Maybach engines, each 250 h.p. — maximum speed was 66 m.p.h..

THE STRAFED ZEPP. L48. June 17 1917. No. 7   J. S. Waddell. Photo. Leiston

L 48 ZEPPELIN

The L 48 was captained by Franz Eichler, and also on board was a very important guest — Captain Victor Schutze, Commodore of the North Sea Airship Division. Both men died along with all but two of the crew.

THE STRAFED ZEPP. L48, June 17, 1917. No. 10   J. S. Waddell, Photo, Leiston

L 48 ZEPPELIN

At Ordfordness three pilots, Captain Saundby flying a DH2, Second Lieutenant Holder flying a FE2b, and Lieutenant Watkins flying a BEl2, climbed to intercept. The L 48 was flying at 15,000 feet, causing the aircraft to struggle to get within firing range. Finally the pilots opened fire from different angles. The L 48 broke into a huge V and fell to earth, 'blazing with a tremendous roar.'

The remains fell in a field near Holly Tree Farm at Theberton, Leiston, Suffolk.

THE STRAFED ZEPP. L48, June 17, 1917.   No. 11   J. S. Waddell, Photo. Leiston

L 48 ZEPPELIN

All three pilots hit the Zeppelin, but it was difficult to credit one man with the decisive shot. Today, Saundby, Holder and Watkins all deserve equal credit. But at the time each was credited with the downing of the L 48. In 1917 each pilot had his own champion.

L 48 ZEPPELIN

There were two survivors in the forward gondola. The other crew members were buried in Theberton churchyard and later interred at Cannock Chase, Staffordshire, when the German cemetery was dedicated. Today the field is known as Zeppelin Field.

The Strafed Zepp. L48. June 17, 1917. No. 16.   J. S. Waddell, Photo. Leiston

French Aeroplane attacking German Airship
The Airship is turning to retreat and is consequently for-shortened.
The Aeroplane being so much nearer the Camera makes the Airship
look small in proportion. This is the first photograph taken in mid-air
of an aeroplane attack

DAILY MAIL · COPYRIGHT

£100 PRIZE WINNER

FRENCH ATTACK

An unidentified Zeppelin returning to base is being attacked by a Vickers FB5. The Darracq Company built 99 of these aircraft between May 1915 and June 1916. It was a very slow climb to the required height for these aircraft, in order to reach the Zeppelins.

LETHAL CARGO
This collection of incendiary bombs was dropped on Bury St Edmunds, Suffolk, during various raids in 1915/1916. There were 41 gathered in this yard, the wire handles by which they were dropped being easily seen.

WE ARE EXPECTING TO LEAVE
THIS PLACE AT ANY MOMENT

BOMBS AWAY, p.u. 1915
A comic/serious card with two Germans wearing their pickelhaubes dropping a fused bomb on our lovers. Posted to Weymouth in 1915, the year when the first raids by Zeppelins were made on England.

BOMBS AWAY, p.u. 12 July 1917
A Zeppelin bombing a 'large shell
factory' somewhere in England,
supposedly from a German official
report in 1916. Posted from Hatfield
Peverel on 12 July 1917 to a Private
F. Fleming — No.3 Company A.O.C.,
A.P.O. No.1., B.E.F. France.

"MORE FRIGHTFULNESS"
"Our gallant airmen successfully bombed a huge shell factory."
Extract from German official report.
Sanctioned by Censor, Press Bureau, October 10th, 1916.

Kapitänleutnant Stabbert
Kommandant eines Marine-Luftschiffes
574

FRANZ STABBERT

Stabbert's earliest raid was on 31 January 1916 in the L 20 — bombing Loughborough and Burton-on-Trent. Then, in May he bombed Craig Castle in Scotland. On returning to base, the L 20 crashed near Sandnaes. Nine of the crew and Stabbert were interned by the Norwegians — six months later they escaped. His next Zeppelin, the L 44, raided Harwich on 24 May 1917. He made other raids in August and September of that year.

CAMPAGNE DE 1914-1916

Un avion boche descendu par une section d'autos-canons.
A german airship brought down by a section of motor car guns.

ND. Phot.

FRANZ STABBERT

On 19 October 1917 the L44 raided Bedfordshire, St Albans, Gravesend and Maidstone. Returning home over France, mobile anti-aircraft guns hit the L 44 and it crashed near Chenevières. All the crew were killed, including its Captain Stabbert.

SL 5 ZEPPELIN

This dramatic card shows the Schutte-Lanz Zeppelin, SL5, after it had made a forced landing in a gale on the eastern front on 5 July 1915. It had a very short life. Built in 1915, it was 502 feet in length and powered by 4 Maybach engines, each 210 h.p.. It was an army Zeppelin, and troops can be seen helping to contain the crippled 'Zepp'.

LEEFE ROBINSON V.C.

Leefe Robinson was born on 14 July 1895 at Co-org State in Southern India. After a happy childhood in a large family, 'Robbie' came to England in 1909 for his education. Later he joined the Officers' Training Corps, swiftly gaining the rank of sergeant. On 4 August 1914, when 19, he entered Sandhurst, hoping to gain a commission in the Indian Army. He was gazetted on 16 December to the Worcester Regiment, joining the Fifth Militia Battalion.

166.F.   LIEUT. W. L. ROBINSON. V.C.   BEAGLES POSTCARDS.
THE INTREPID AIRMAN WHO, ON 3RD SEPT. 1916, ATTACKED AN ENEMY ZEPPELIN (L.21) IN MID-AIR, BRINGING IT DOWN TO THE EARTH IN FLAMES.

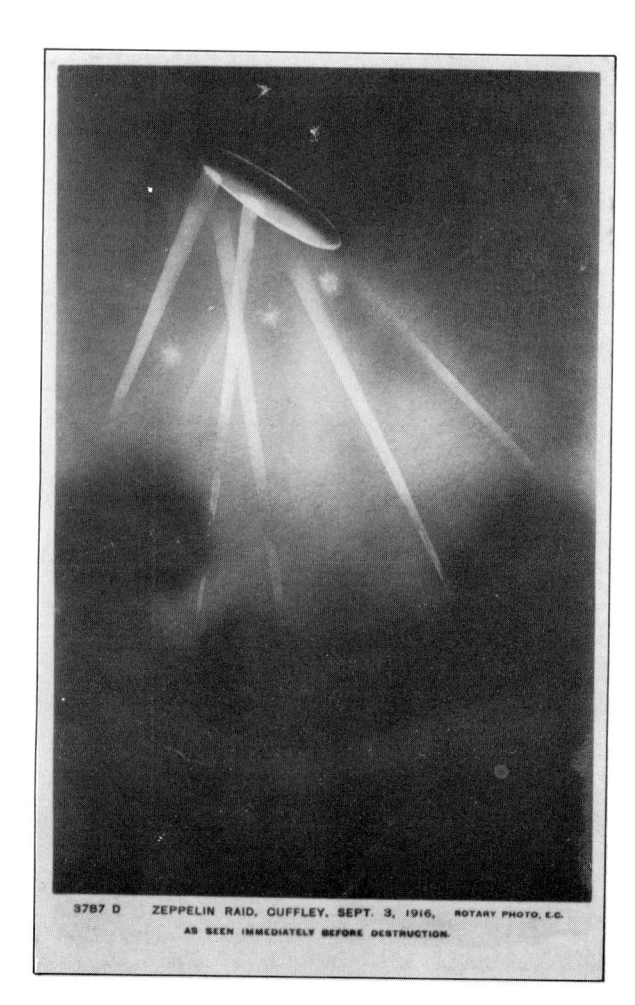

3787 D   ZEPPELIN RAID, CUFFLEY, SEPT. 3, 1916,   ROTARY PHOTO, E.C.
AS SEEN IMMEDIATELY BEFORE DESTRUCTION.

LEEFE ROBINSON V.C.

Second Lieutenant Robinson found army life tedious and boring. His days were spent as an orderly officer, trying to interest his men with the daily chores. His need for excitement and action made him apply for a transfer to the Royal Flying Corps. Later he was told to report to No.4 Squadron at St Omer, France on 29 March 1915. His squadron life took on a more demanding role when he became an observer flying in a BE2c. Then, in May 1915, he was wounded by shrapnel and was invalided home, by now wishing and hoping to become a pilot.

LEEFE ROBINSON V.C.

After convalescent leave he reported to Farnborough for pilot tuition in June 1915. Flying a Farman 'Longhorn', he qualified after 3 hours 50 minutes, and in September, Flying Officer Robinson joined No. 19 Squadron at Castle Bromwich. He was an excellent pilot and loved everything about flying. Most of it was done with the Home Defence Squadron — ferrying aircraft. During all this time Zeppelins were making repeated raids on England, and there was a need for night-flying pilots to intercept the raiders.

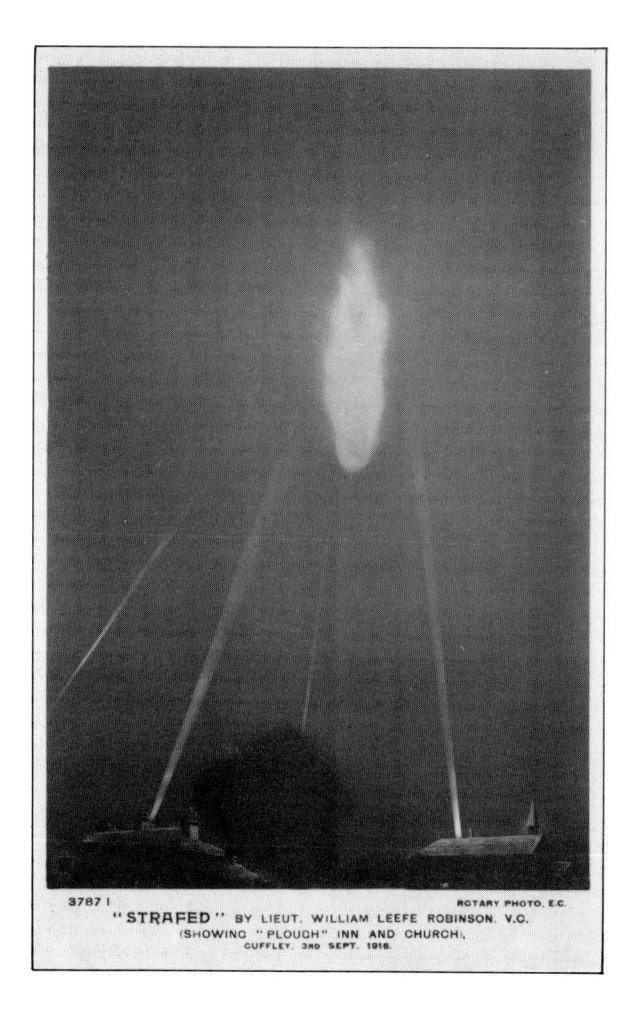

37871   ROTARY PHOTO. E.C.
"STRAFED" BY LIEUT. WILLIAM LEEFE ROBINSON. V.C.
(SHOWING "PLOUGH" INN AND CHURCH).
CUFFLEY. 3RD SEPT. 1916.

59

L21, brought down in flames at Cuffley, Herts, by
FLIGHT-LIEUT. ROBINSON V.C. Sept. 3rd. 1916

LEEFE ROBINSON V.C.,
p.u. 27 SEPTEMBER 1916
The second of September 1916 found sixteen Zeppelins being prepared for a raid on London. At Sutton's Farm airfield, Robinson and other pilots were waiting with their BE2c aircraft. It took Robbie over an hour to reach 12,000 feet; the SL 11 Zeppelin was picked out by searchlights and the attack began. The BE2c made passes along both sides of the 650-feet-long airship. Concentrating his fire on the rear part of the 'Zepp', he watched as Wilhelm Schramm and his crew were engulfed in flames. This card was posted from London on 27 September 1916 to Hordle, Hampshire. Down the right-hand side the writer states, 'This is very good I saw it all.'

3787 A   ZEPPELIN WRECK, CUFFLEY, 3RD SEPT., 1916.   ROTARY PHOTO, E.C.
GENERAL VIEW OF WRECKAGE.

LEEFE ROBINSON V.C.

The SL 11 crashed behind the Plough Inn at Cuffley, Hertfordshire. War Office officials who were on the scene during the day were hindered from their investigations of the wreckage. Armed troops were mounted to keep souvenir hunters away, but not before many pieces of the wreck had been taken away.

3787 B   THE WRECKED ZEPPELIN, CUFFLEY, 3RD SEPT., 1916.   ROTARY PHOTO, E.C.

LEEFE ROBINSON V.C.

Three men of the Royal Flying Corps seen here dismantling the wreckage, with Navy and Army personnel searching the debris, seeking further information on Zeppelin technology.

Lieut. W. L. Robinson, V.C. Cheered by his fellow airmen after destroying Zeppelin, Sept. 3rd 1916.

LEEFE ROBINSON V.C.

On his return to the squadron he was hailed as a hero. To the public he was a young, dashing saviour who had slayed the enemy. Robbie found it hard to travel about without being recognised and mobbed. Soon after the destruction of the Zeppelin, his face was in all the papers, magazines and on postcards. There were even train excursions from Kings Cross to Cuffley to see the scene of this heroic deed.

Lieut. William Leefe Robinson, V.C.

LEEFE ROBINSON V.C.

Lieutenant Robinson was summoned on 8 September to Windsor, where King George V pinned the Victoria Cross to his tunic. Both men spent several minutes talking of the events of the last few days. On 5 April 1917 Robinson took off from La Bellevue on a patrol. Later that day his flight was attacked by Albatross D lll's over Arras. In the combat Robinson was forced to land behind German lines. He and his observer were taken prisoner. After over a year of changing camps and attempting to escape, his time in Germany came to an end. On 14 December 1918 he was repatriated. His health had deteriorated and he had no resistance to an influenza epidemic, and on 31 December 1918 he died, aged 23 years.

FARNBOROUGH, c.1917

A multi-view postcard showing scenes of: the Common with an observation balloon, the Aircraft Factory and hangers, Church Parade with band and airmen, the Queen's Avenue with a biplane overhead, and centre, the Queen's Hotel. The card is dated 1917.

VICTORY EXHIBITION

This observation balloon was part of the Royal Air Force section display at the Crystal Palace Victory Exhibition shortly after the First World War. The balloon was tethered, but, from eye-witness reports, it used to move around, or at least sway. The Crystal Palace burnt down in 1936.

ZEPPELIN HUMOUR,
p.u. 15 November 1915

Many of the cards depicted people and animals hiding under things; umbrellas were a favourite hiding place. This card was posted to a local address in Keighley, Yorkshire on 15 November 1915. It was published by Bamforths of Holmfirth who produced a tremendous number of comic cards around this period.

ZEPPELIN HUMOUR

Another umbrella, this time for two, drawn by the famous artist Donald McGill. Posted from Watford to Wellingborough, Northamptonshire on 3 August 1915. Part of the message reads, 'We went over to a soldiers' camp on Sunday, & had tea.'

ZEPPELIN HUMOUR, c.1915
Many comic cards of this type were published during 1915 and 1916 to boost the morale of civilians when the Zeppelin raids were at their peak. This disgruntled old gentleman is disturbed by the throb of the engines.
The artist is Lawrence Colborne.

Hark ! I hear a Zeppelin !

You did startle me ; I thought
you were a Zeppelin.

ZEPPELIN HUMOUR,
p.u. 25 March 1916

The right shape, but far too small. This card was sent from Halton Camp, Buckinghamshire to Rushden, Northamptonshire on 25 March 1916. Halton Camp was a Royal Flying Corps establishment at this time, training and turning out skilled airmen in many trades. The camp continues to this day under the Royal Air Force.

BODENSEE

After the end of the war the Deutsche Luftschiffahrts — Aktien — Gesellschaft (DELAG) Company was enlarged, with the intention of expanding its commercial activities. The Bodensee, or LZ120, was built in 1919. Only 400 feet in length, it was powered by 4 Maybach engines, each 260 h.p.. During its time with DELAG it made 103 flights in 81 days, carrying 2,253 passengers. However, the Allies demanded the DELAG fleet as reparation, and the Bodensee was sent to Italy, where later it was broken up.

H.M. AIRSHIP "R 34."
Model of the rigid airship built by Messrs. Beardmore during the latter part of the war and the first lighter-than-air craft to achieve a direct trans-Atlantic flight. Overall length, 640 ft.; diameter, 79 ft.; capacity, 1,959,000 cub. ft.; maximum speed, 52 knots; cruising speed, 42 knots; gross lift, 59 tons.

The Science Museum, London. No. 57

R34 AIRSHIP

On 2 July 1919 the R34 left East Fortune, on the Firth of Forth, on the first transatlantic round trip. She was 643 feet in length and was powered by five Sunbeam engines, each engine 250 h.p.. Maximum speed was 62 m.p.h. She carried nearly 16 tons of fuel, three tons of water ballast and four tons of crew and gear. Her Captain, Major G.H. Scott, (later to perish in the R101) landed her at Mineola, Long Island, 108 hours 12 minutes later, having flown 3,600 miles. She landed with only 40 minutes of fuel left. The return flight on 10 July took 75 hours 3 minutes. She landed at Pulham, Norfolk having been diverted from East Fortune. The R34 broke up in 1921 at Howden, Yorkshire due to the lack of proper mooring facilities and adverse weather conditions. In all she had flown about 500 hours.

R80 AIRSHIP, p.u. 30 July 1920

Designed by Barnes Wallis and built by Vickers, the R80 was launched at Walney Island, Barrow-in-Furness, in July 1920. A streamlined airship, even the control car was shaped like the cockpit of a modern airliner. During her short career she made four training flights with the United States Navy. It was an efficient airship, but too small for any commercial programme. She was 535 feet in length and powered by 4 Wolseley-Maybach engines. On 21 September 1921 she made her last flight, from Howden to Pulham, Norfolk, and then in 1925 she was dismantled as an economy measure. This card was posted from Barrow on 30 July 1920 to Northampton. Part of the message reads, 'this is the new airship launched the other week'.

R101

The R101 was launched at Cardington, Bedfordshire on 12 October 1929. She was 777 feet long, the mast to which she was moored was 202 feet high and 70 feet in diameter at its base. Its maiden flight lasted over 5 hours, and its longest flight prior to its last had been a 'flag waving' trip lasting 30 hours around most of Britain. But before and after these trips problems had been found. Engine faults, faulty gas valves, chafing gas bags and the general overweight of the whole thing. Modifications were made, the most serious being the cutting in half of the airship to insert an extra section to carry more gas bags to increase lift. The Government and Lord Thomson (Secretary of State for Air) had their eyes and hopes set on using the R101 to take VIPs to India for the Conference of Empire Ministers. But it was not an over-confident crew and ground staff. Many felt uneasy at the prospect of such a long journey.

R101

At 6.36pm on 4 October 1930, the R101 slipped her mooring mast in damp weather, circled Bedford as was tradition, then turned south on her 5,000 mile flight to India. On board was a crew of 48, including technical advisers, and 6 passengers including Lord Thomson and Sir Sefton Brancker (Director of Civil Aviation). The airship took over one hour to reach London, crossing the Channel near Hastings at around 9.35pm. Loss of power in one engine, heavy rain on the fabric adding to the weight and the problem of porpoising were causing some concern. She tried to maintain a height of 1,500 feet over the French countryside. It was nearly ten past two on the morning of 5 October that the great airship ploughed into the ground near Beauvais — bursting into flames and leaving a skeletal mass of girders. Forty eight men perished; there were six survivors.

R101

The country was stunned. A funeral with full state honours was prepared. The French government had made special arrangements for the bodies to be carried to Boulogne. There, two British destroyers, the Tempest and Tribune, carried the coffins to Dover. From there a special train conveyed the 'valiant dead' to Victoria. King George V had given permission for Westminster Hall to be used for the lying-in-state on 10 October. Raised on a dais of purple were the forty-eight coffins, each draped with the union flag and surrounded by masses of flowers and wreaths, and guarded by seven airmen with reversed rifles.

R101

The next day the coffins left Westminster and were borne to Euston Station for their journey to Bedford. All along the route stood bare-headed people, paying homage as the train approached the hushed and silent town. Royal Air Force Crossley tenders, each carrying two coffins, slowly made their way from Bedford Station. There was an estimated crowd of 75,000 people lining the 3 mile route as the procession neared Cardington village.

R101

The bell in St Mary's church had been tolling for half an hour when the first tender came into view. The coffins were carried into the common grave, laid in four rows twelve deep, still draped with their flags. The Bishop of St Albans, the Vicar of Cardington, the Chaplain in Chief of the RAF, the Roman Catholic Bishop and Nonconformist ministers took part in the committal service. On the word of command three volleys rang out, then the Last Post and Reveille sounded. The R101 victims were laid to rest among their own people.

Memorial to R101. Heroes, Cardington Cemetery. 137723

R101

After some time the grave site was partially grassed over and in the centre an area was covered with stone slabs. On top of these a large white rectangular stone memorial was placed. On its sides were the names of the 48 buried beneath, and at one end, the badge of the Royal Air Force, in bold relief.

The R101 Memorial
Cardington Church.

137422

R101

In St Peter's church, Cardington, is the partially charred ensign that flew from the R101. It is preserved in a glass-fronted case. Beneath is a memorial to those that died, their names all listed. Opposite the scene depicted on this card is a large photograph (also framed) of the airship moored at its mast, as she had been only a few hundred yards from this church.

R101

This In Memorium card was produced after the fatal crash by a firm in London. It gives all the details of the disaster and also names all those that died. The card folds into an area 4½ x 3 inches, printing on all four faces.

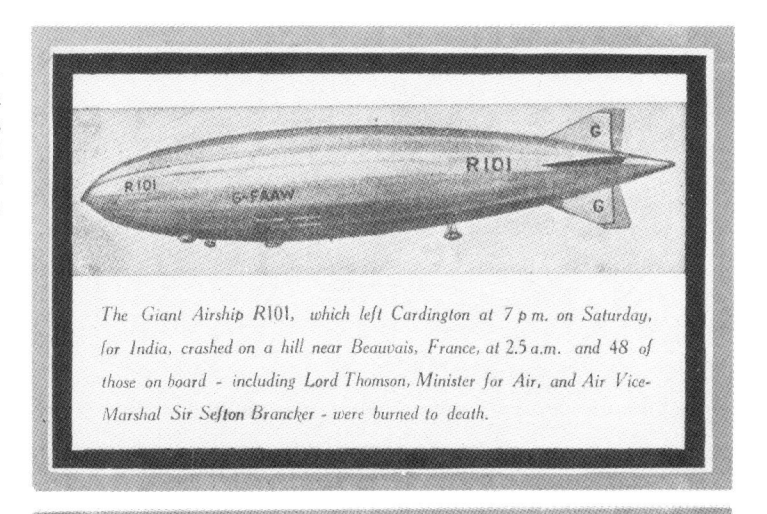

The Giant Airship R101, which left Cardington at 7 p.m. on Saturday, for India, crashed on a hill near Beauvais, France, at 2.5 a.m. and 48 of those on board - including Lord Thomson, Minister for Air, and Air Vice-Marshal Sir Sefton Brancker - were burned to death.

In Sacred Memory of

THE OFFICERS, CREW and PASSENGERS of the British Airship R.101

which Crashed in France on

Sunday Morning OCT. 5th 1930.

with a loss of 48 lives including Lord Thomson, Air Minister and Sir Sefton Brancker, Air Vice Marshal.

May Their Souls Rest In Peace.

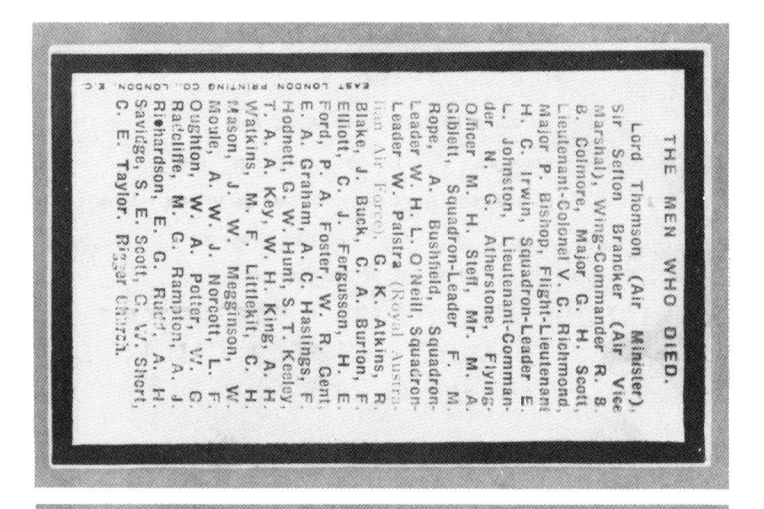

THE MEN WHO DIED.

Lord Thomson (Air Minister).
Sir Sefton Brancker (Air Vice-Marshal), Wing-Commander R. B. Colmore, Major G. H. Scott, Lieutenant-Colonel V. C. Richmond, Major P. Bishop, Flight-Lieutenant H. C. Irwin, Squadron-Leader E. L. Johnston, Lieutenant-Commander N. G. Atherstone, Flying Officer M. H. Steff, Mr. M. A. Giblett, Squadron-Leader F. M. Rope, A. Bushfield, Squadron-Leader W. H. L. O'Neill, Squadron-Leader W. Palstra (Royal Australian Air Force), G. K. Atkins, R. Blake, J. Buck, C. A. Burton, F. Elliott, C. J. Fergusson, H. E. Ford, P. A. Foster, W. R. Gent, E. A. Graham, A. C. Hastings, F. Hodnett, G. W. Hunt, S. T. Keeley, T. A. A. Key, W. H. King, A. H. Watkins, M. F. Littlekit, C. H. Mason, J. W. Megginson, W. Moule, A. W. J. Norcott, L. F. Oughton, W. A. Potter, W. G. Radcliffe, M. G. Rampton, A. J. Richardson, E. G. Rush, A. H. Savidge, S. E. Scott, C. W. Short, C. E. Taylor, Rigger Church.

EAST LONDON PRINTING CO., LONDON, E.C.

Thy Will be done

The back and front faces of the card: the outer edges are coloured silver with an inner edge of black. This colour scheme is the same on all four faces.

GRAF ZEPPELIN I

The LZ127, or Graf Zeppelin I, was launched in 1928. She made many flights to all parts of the world. Her trans-Siberian flight was a huge success, and was the first non-stop flight over Russia. Her pilot, Hugo Eckener, had carried on the work of Count Zeppelin after his death. In 1929 she carried out a survey of the Arctic, and in 1930 pioneered a regular service to South America. The Graf Zeppelin was broken up in 1940, after many years service. This card shows her flying over Cologne Cathedral.

L.Z. 127. "Graf Zeppelin" über dem Kölner Dom

Luftschiff L. Z. 127 „Graf Zeppelin" nach seiner Landung

GRAF ZEPPELIN

This excellent card shows the enormous size of the Graf Zeppelin; its length was 775 feet, its diameter 100 feet and it stood 113 feet high. The public were often asked to man-handle these giants, when the airships were pulled around or nudged into a different position after landing. Handgrips can be seen on the side of the cabin, each one being held by a willing helper.

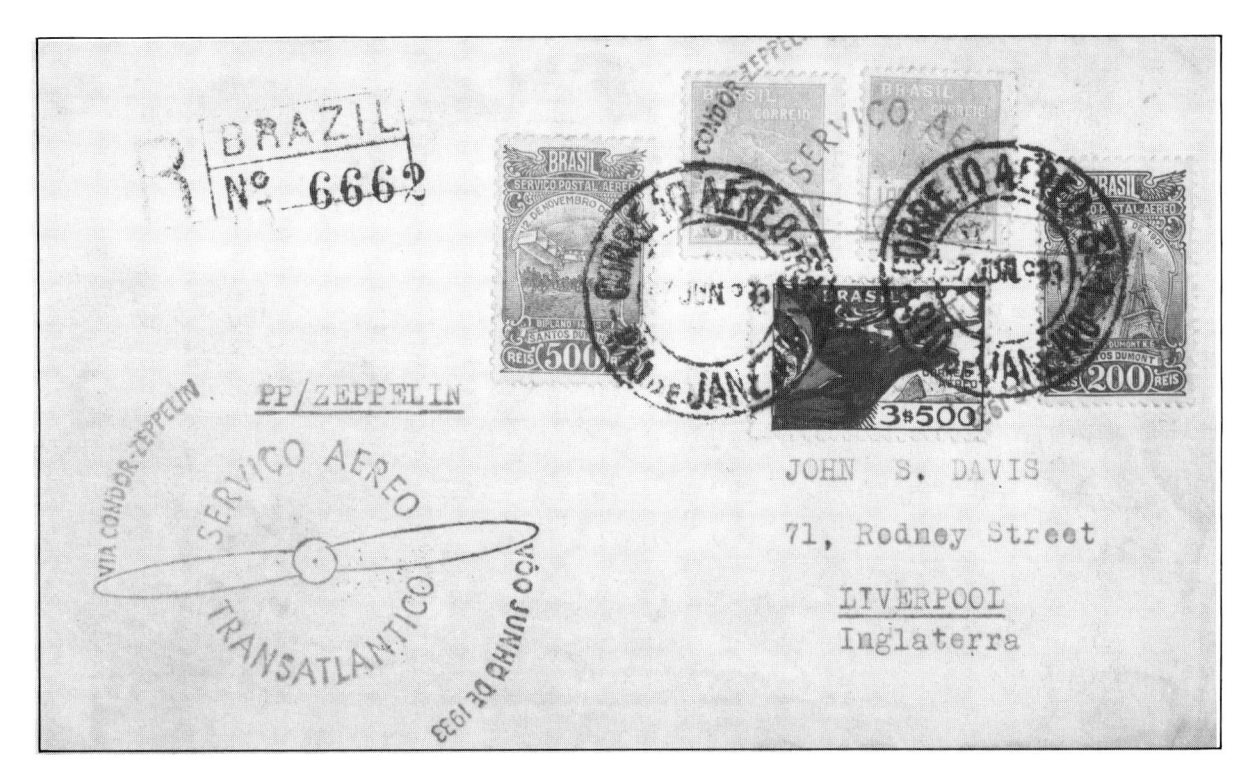

ZEPPELIN MAIL

Two examples showing the types, (although there are more) of cachets used on the Atlantic runs. This registered letter flew from Rio de Janeiro, in June 1933, to Liverpool, England. On the reverse of this envelope is a green Friedrichshafen (Bodensee) handstamp. This was the second 1933 South American flight.

ZEPPELIN MAIL

This envelope was flown on a 1939 flight from London, via Germany, to Buenos Aires, Argentina. Showing the cachet, in red, of an airship and aeroplane coming out of clouds over the sea — this type of mark was widely used and is the more common.

SHENANDOAH

In 1919 the United States Secretary of the Navy authorised the building of two rigid airships. One was the ZRI, known as the Shenandoah. She made her first flight on 4 September 1923, and remained in service until September 1925, when it was caught in a storm over Ohio and broke in two. Out of a crew of 43, the survivors numbered 29. With a length of 680 feet, and diameter of 78 feet, its envelope contained 20 gas bags. She was powered by 6 Packard engines each 357 h.p. with a maximum speed of 60 m.p.h.

LZ129 HINDENBURG

The dramatic picture shows the Hindenburg catching fire as she came to her mooring mast at Lakehurst, New Jersey on 6 May 1937. It had just completed its 63rd flight. Of these, 37 had been ocean crossings. Launched in 1936, it was 804 feet in length, with a diameter of 135 feet. The Hindenburg was powered by 4 Mercedes Benz diesel engines, each 1200 h.p., with a cruising speed of 78 m.p.h.. Of the crew and passengers, 35 died and 62 were saved. This tragedy really saw the end of the Zeppelins and this type of transport. Only the Graf Zeppelin II outlived the Hindenburg, and she made her last flight on 20 August 1939 and was broken up at Frankfurt in March 1940.